A [illegible]
of the Tropical
Rain Forest

Written by Judy Ling
Photography by Datuk Ling Beng Sung

The World's Tropical Rain Forests

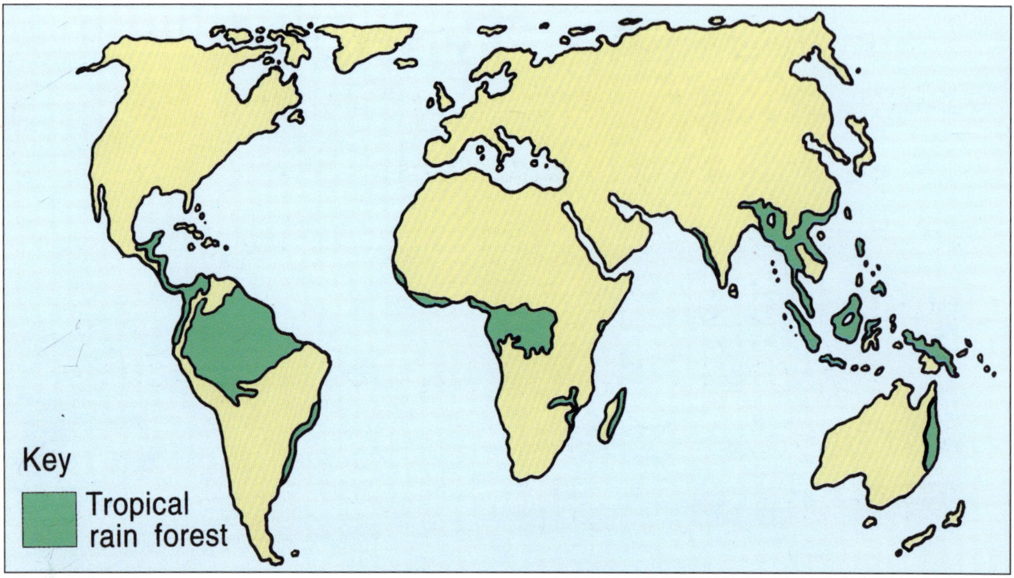

The world's tropical rain forests
are found where the climate
is always warm
and there is plenty of rain.
These rain forests cover only
a small part of the earth,
yet they are home to most
of the world's plants and animals.
This book is about just a few
of those amazing animals.

An Acrobat of the Treetops

I am the acrobat of the treetops.
With my long arms and fingers,
I can swing my way from branch
to branch faster than any other
animal in the rain forest.
I have thick, shaggy fur, but my face
and hands are bare.
What am I?

I am a gibbon.

Gibbons are very noisy.
As soon as they wake up
in the morning, they hoot and call
a warning to tell other animals
to keep away.
Gibbons' favorite foods are fruit,
leaves, flowers, insects, and birds.

A Bird with a Dangerous Bill

The most interesting thing about me
is my long, brightly colored bill.
My bill is dangerous enough
to scare away monkeys
and to fight off snakes.
I can also use it to pluck and peel
the fruit I love to eat.
What am I?

I am a hornbill.

Hornbills have a strange way of nesting. The mother bird makes a nest in the hole of a tree. After she has laid her eggs, she is walled up inside the nest and fed through a tiny hole by the male bird. When the chicks have hatched, the mother breaks out of the nest, so she can help feed her new family.

A Reptile That Lives Near Water

I live in the rain forest,
making my home near water.
I am a reptile, like snakes and turtles.
Because I am cold-blooded, I need
the warmth of the sun to keep
my body temperature warm.
I have dry, scaly skin, clawed toes,
and a tail.
What am I?

I am a lizard called
the oriental water dragon.

The oriental water dragon lives in trees that are close to water. These lizards are good swimmers and divers. They eat insects, spiders, frogs, smaller lizards, birds, and sometimes fruit.

A Miniature Deer?

Although I have hoofs, like a horse or a deer, I am a tiny animal, even smaller than a cat.
Because I sleep during the day and come out at night, you won't often see me in the wild.
What am I?

I am a mouse-deer.

The mouse-deer is not really a deer, even though it has hoofs.
It can run almost as soon as it is born.
Its tiny size helps it to run swiftly through the thick forest to escape its predators.

A Creeping Forest Dweller

I am related to monkeys.
I have huge eyes, and my arms
and legs are long and skinny.
My strong fingers and toes can grip
tree branches very tightly.
I creep along the branches very slowly —
maybe that's how I got my name.
What am I?

I am a slow loris.

The slow loris is a nocturnal animal. It sleeps during the day and comes out at night. It has huge eyes and an excellent sense of smell. It eats plants as well as insects and small reptiles.

A Wonderful Singer

I have beautiful feathers, and my tail can be over three feet long.
People don't often see me in the wild, but my lovely song can be heard from far away.
I eat fruit, insects, nuts, and seeds.
What am I?

I am the great Argus pheasant.

When a male great Argus pheasant
is looking for a mate, he finds
a clearing in the forest.
Here the male bird dances
and shows off his feathers to try
to attract the attention of a female.

A Clinging Tree Climber

I am small and green—
the same color as the leaves
of the trees where I live.
With my bulging eyes, I can see
in all directions.
I have strong legs that help me leap
and sticky pads on my feet
that help me cling to trees and leaves.
What am I?

I am a tree frog.

This is the green tree frog. Because it is almost the same color as its background, its predators find the green tree frog difficult to see.

The World's Largest Parrot

I am the largest member
of the parrot family.
I have a long tail and bright feathers
of blue, red, yellow, or green.
Whenever I see a stranger, I shriek out
loudly to warn the other animals.
What am I?

I am a macaw.

The macaw has a powerful beak for cracking nuts and seeds. Its beak is even strong enough to break open the Brazil nuts it loves to eat.

Man of the Forest

My name means "man of the forest."
I am the largest animal
that lives in trees.
My long, powerful arms help me
swing from one tree to another.
I spend most of the day eating,
and at night I make a nest to sleep in.
What am I?

I am an orangutan.

People are the orangutan's most feared enemies. People have destroyed much of the rain forest home of orangutans, and many of these animals have been killed. In many countries where orangutans live it is against the law to kill or trap orangutans.

A Silent Night Hunter

I usually sleep during the day
and hunt for food at night.
I have large, round eyes
and a sharp, hooked beak.
My soft, fluffy feathers help me
fly quietly so that my prey
doesn't hear me coming.
I eat small birds, rodents, and insects.
What am I?

I am an owl.

This bay owl lives in the rain forests of Southeast Asia.

Did you know . . . ?

Like people, owls have two eyes pointing forward. This means they can watch the same things with both eyes at the same time. This gives the owl extremely good eyesight to help it catch its prey.

The Future of Tropical Rain Forests

Today, much of the tropical rain forest has been destroyed, and this destruction is still going on. Many animals have lost their homes, and some species have died out. Others are now in danger of extinction. Let's hope it is not too late to save the animals of the tropical rain forest.

Index